Fluid Power
Educational
Series

Accumulators in Hydraulic Systems

(In the English Units)

Joji Parambath

Accumulators in Hydraulic Systems
(In the English Units)

ISBN: 9798653881503

https://jojibooks.com

First Edition – 2020
Revised Edition - 2021
Revised Edition - 2026

Disclaimer of Liability

The contents of this book have been checked for accuracy. Since deviations cannot be precluded entirely, we cannot guarantee full agreement. Only qualified personnel should be allowed to install and work hydraulic equipment. Qualified persons are defined as persons who are authorized to commission, ground, and tag circuits, equipment, and systems following established safety practices and standards.

Table of Contents

Preface

Hydraulic accumulators are key devices in hydraulic systems, performing control functions, but their functions, construction, and control circuits can be complex for untrained professionals.

This book provides essential technical information on accumulators, covering functions, classification, construction (piston, diaphragm, and bladder types), pre-charging, safety, and applications, along with basic circuits and sizing examples in English units.

Maintenance and specifications are also included.

The content is organized from simple to complex for quick understanding, aimed at newcomers and untrained professionals.

Additional fluid power topics are available in the author's other textbooks, listed at the end and on https://jojibooks.com.

Enjoy reading the book.

Your feedback is most welcome.

JOJI Parambath

Chapter 1 | Functions of Hydraulic Accumulators

An accumulator is a device that absorbs shock pressures and stores energy in a hydraulic system. It mainly consists of a vessel that holds hydraulic fluid under pressure using a raised weight, a spring, or a volume of compressed gas. Thus, potential energy can be stored in the accumulator when the system pressure exceeds the accumulator's pressure. The accumulator can release stored energy back into the system to perform a useful hydraulic task when the system pressure falls below the accumulator's pressure. Figure 1.1 shows a realistic view of a gas-charged accumulator with a safety control block.

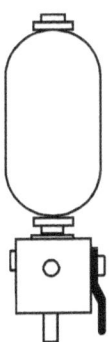

Figure 1.1 | A hydraulic accumulator with a safety shut-off valve

Hydraulic accumulators play a vital role in machinery, enhancing performance, boosting energy efficiency, and reducing noise levels. This book guides you through different aspects of these important components, including their functions, construction, types, safety tips, circuits, maintenance, and sizing. The key functions of accumulators are explained in the upcoming sections.

Generation of Shocks in Hydraulic Systems

Shock pressures develop in a hydraulic system when flow is abruptly blocked or the flow direction is suddenly changed, as when the associated directional valve is shifted. Shocks may also develop from the jerkiness of the load attached to a cylinder or motor in the system. Shocks may also result from external mechanical forces.

Shock Absorbing Function of Hydraulic Accumulators

Two circuits shown in Figure 1.2 demonstrate the effect of shock pressures on a hydraulic system with a pump, valve, actuator, and interconnecting lines, and the elimination of shock pressures by adding an accumulator to the system.

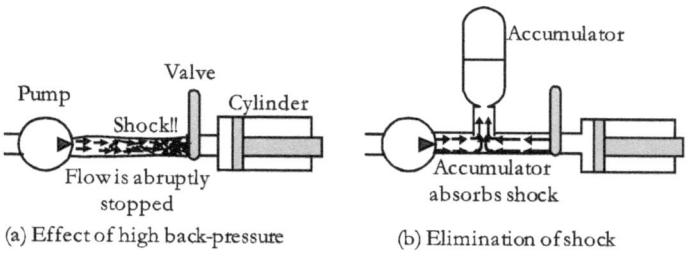

(a) Effect of high back-pressure

(b) Elimination of shock

Figure 1.2 | Schematic diagrams illustrating the development of shock pressure in a hydraulic system and its elimination

Figure 1.2(a), shows the movement of a column of high-energy fluid through the hydraulic circuit without an accumulator. When the flow is blocked abruptly, for example, with the rapid closure of the valve, pressure waves are developed in the system, which travel back and forth through the system. The resulting hydraulic shock, or 'water hammer' as it is sometimes called, and the vibration can develop peak pressures several times the normal working pressure. It may cause severe damage to the hydraulic system's components. For example, the shock pressures may rip tubing, blow seals, produce fatigue failures, jar parts, and split the pump housing.

The damaging effects of hydraulic shock can be prevented by installing an appropriately sized accumulator in the system, as illustrated in Figure 1.2(b). The accumulator can absorb the kinetic energy of the moving fluid column and suppress the resulting hydraulic shock. With this action, the system components, such as pumps, valves, hoses, and fittings, are protected from pressure spikes. Note that an accumulator is usually connected as close to the demand point as possible to overcome flow restrictions and drag from long pipe runs.

Other Functions of Hydraulic Accumulators

In addition to storing energy and absorbing shocks, as described in the previous section, accumulators can also perform other functions, such as dampening pressure pulsations, storing energy during periods of low demand, releasing energy during periods of high demand, and compensating for leaks. For example, an accumulator can be used in an application requiring a large volume of hydraulic fluid for a very brief period. By implementing some of these functions, accumulators can reduce pump capacity requirements, system leakage, and noise levels in hydraulic systems. Accumulators can help a hydraulic system reduce energy and maintenance costs and improve performance, efficiency, and reliability.

Pulsation Dampening

Pressure pulsations in a hydraulic system are usually caused by fluctuations in pump delivery, irregularities in fluid flow, thermal variations, or excessive loads. Pressure fluctuations in the system can cause variations in the actuator's speed and reduced system performance. Adding a properly located and correctly sized accumulator to the system cushions pressure pulsations and keeps the system pressure relatively constant. The cushioning of pressure pulsations further reduces noise levels in the system.

Energy Storage and Release

A typical machine cycle during the operation of a hydraulic system consists of extended periods (say, 80% of cycle time) of little flow and short periods (say, 20% of cycle time) of high-volume flow, as shown graphically in Figure 1.3.

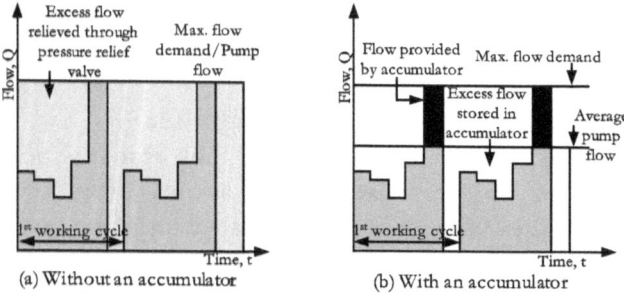

Figure 1.3: | Typical machine cycles in hydraulic systems

If a fixed-displacement pump is used in the system with an intermittent duty cycle to supply the system fluid, then the system pressure relief valve bypasses the fluid for most (say 80%) of the time. If the pump runs continuously, the discharge of the pressurized fluid through the associated pressure relief valve represents a considerable power loss, as shown in Figure 1.3(a).

With the addition of a correctly sized accumulator, energy can be stored during periods of low demand and then added to the pump flow during short periods of high demand. Therefore, the pump unit in the system can be resized to match the average power requirement of the machine's operating cycle, as shown in Figure 1.3(b).

The result is reduced pump drive power. The use of the accumulator also results in a quick response to temporary peak energy demands. A rule of thumb suggests that an accumulator can be incorporated into the system if the pump operates at full load for less than 20% of the time.

Cost Reduction
Using an accumulator in a high-performance hydraulic system with an intermittent duty cycle allows the use of a smaller-capacity pump. As we are aware, an accumulator can store energy during periods of low demand. This energy can be released upon demand and is available for immediate use. In turn, the smaller-capacity pump uses less fluid during system operation, reducing the required reservoir size. The net result is a cost reduction in the system's construction.

Leak Compensation
On hydraulic presses used for molding, bonding, etc., it is necessary to maintain constant pressure on the work during long curing periods. However, pressure changes can occur in such systems due to fluid leakage. A charged hydraulic accumulator can compensate for pressure changes by supplying additional fluid to the system. At the same time, the pump remains unloaded during an extended idle period, making up for the loss of system fluid.

Auxiliary Power Source

Energy stored in a fully charged accumulator can be used in a hydraulic system to meet sudden, high-power demands for a comparatively short time to complete the cycle, or as a standby power source during power failures. It can also be used to independently operate auxiliary or pilot circuits in a hydraulic system when the pump flow is required to perform the main operating movements. Accumulators are not meant for every hydraulic system. They involve cost; therefore, they can be used if they offer an advantage over conventional circuits.

Summary of Basic Accumulator Functions

Table 1.1. gives a summary of basic accumulator functions.

Table 1.1 | Summary of basic functions of accumulators

Function	Description
Shock Absorption and Pulsation Dampening	Accumulators absorb sudden pressure spikes, vibrations, and noise, thereby protecting valves, pumps, and pipes from structural damage.
Energy Storage	Accumulators store hydraulic energy during low-demand periods and release it during high-demand periods, providing a secondary power source .
Leakage and Volume Compensation	Accumulators maintain system pressure over long periods by compensating for minor leaks, preventing the pump from cycling continuously.
Emergency Power Source	In the event of pump failure, accumulators provide stored pressure to complete cycles or return actuators to a safe position.

Chapter 2 | An Overview of Accumulator Types

This chapter presents the classification of accumulators and the constructional features of weight-loaded and spring-loaded hydraulic accumulators. The constructional details of hydro-pneumatic accumulators are presented in subsequent chapters.

Based on construction methods, accumulators are classified into three basic types. They are (1) weight-loaded accumulators, (2) Spring-loaded accumulators, and (3) hydro-pneumatic (gas-charged) accumulators. Weight-loaded types are no longer used in modern hydraulic applications, and spring-loaded types are now virtually confined to small mobile machinery. In modern hydraulic systems, the preferred type is the hydro-pneumatic accumulator. They are further classified as bladder, diaphragm, piston, and bellows types. They are available in a full gamut of sizes, pressure ratings, materials, and port configurations. The classification chart of Figure 2.1 shows all these types and their subtypes.

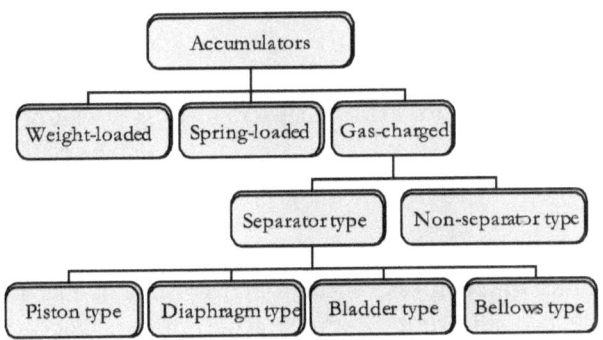

Figure 2.1 | Classification of hydraulic accumulators

General Constructional Features of Accumulators

Figure 2.2 shows the simplified cutaway views of the different types of hydraulic accumulators. They are not complete representations of accumulators, but they illustrate the general constructional features of accumulators for initial learning.

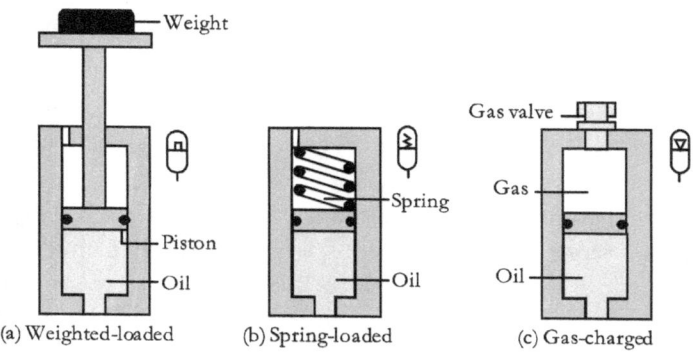

Figure 2.2 | Cross-sectional views of different types of hydraulic accumulators

An accumulator usually consists of two chambers separated by a piston, diaphragm, or bladder (bag). One chamber is designed to admit fluid from the associated system, and the second chamber is used to maintain a weight, spring, or volume of pressurized gas. Hydraulic energy is stored when the system fluid under pressure acts against a weight-loaded or spring-charged piston, or a gas-charged piston, diaphragm, or bag.

Accumulator Symbols

The standard symbols, shown in Figure 2.3, are used to describe the different types of accumulators according to ISO 1219. The symbols are similar to those in ANSI.

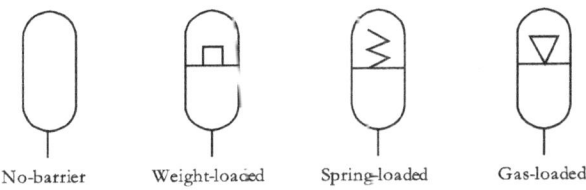

No-barrier   Weight-loaded   Spring-loaded   Gas-loaded

Figure 2.3 | Symbolic representations of hydraulic
accumulators

Weight-loaded Accumulator

The weight-loaded hydraulic accumulator consists of a
vertically-mounted, thick-walled steel cylinder with a piston.
A deadweight or a series of dead weights is placed on the
top of the piston, as shown in Figure 2.4. A correctly-sized
port is provided for the hydraulic connection. The non-
pressure side usually has a drain port (not shown) to relieve
any leaking fluid and prevent backpressure on the piston.

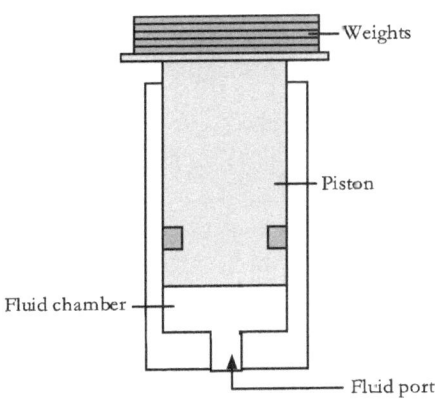

Figure 2.4 | A weight-loaded accumulator

The piston, along with the dead weight, is raised when the
fluid under pressure acts against it. At the same time,

9

gravity exerts a downward force on the piston. As the system pressure drops, the fluid from the accumulator is forced out into the hydraulic circuit by the weight.

An advantage of the weight-loaded accumulator is that it maintains a constant fluid pressure across the unit's output volume. It is also capable of supplying a large volume of fluid under high pressure. Accumulators of this type are used in large forging and molding presses. The weight-loaded accumulators are bulky, cumbersome, and expensive. Moreover, their portability is not feasible.

Spring-loaded Accumulator

The spring-loaded accumulator, as shown in Figure 2.5, consists of a cylinder body, movable piston, and spring. The piston in the chamber is pre-loaded with a spring. The non-pressure side usually has a drain port (not shown).

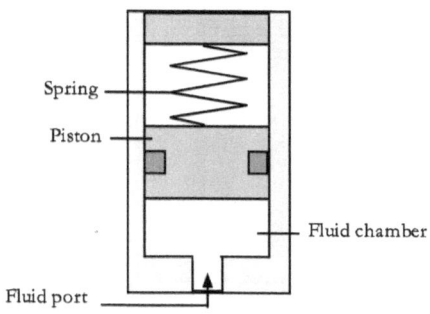

Figure 2.5 | A spring-loaded accumulator

As pressurized fluid enters the chamber from the associated hydraulic circuit, the spring compresses. The compressed spring always acts against the piston. As system pressure decreases, the fluid from the accumulator is forced into the circuit by the charged spring.

Spring-loaded accumulators are usually smaller and less expensive than weight-loaded accumulators. Their mounting is easy, and they can be fitted directly to the power unit. The pressure generated by this type of accumulator depends on the spring's size and preload.

The accumulator pressure peaks when the spring is fully compressed and drops to its minimum as the spring approaches its free length. Therefore, the pressure does not remain constant across the spring-loaded accumulator's volume output. Spring-loaded accumulators are typically suitable for low-pressure, low-volume hydraulic applications but not for high-cycle-rate applications.

Gas-charged Accumulators
The low compressibility of the hydraulic fluid makes it difficult to store hydraulic energy in small volumes, but it can transfer a considerable amount of force to a load surface. On the other hand, gas is a highly compressible medium that can store large amounts of energy in a small volume. Gas-charged accumulators leverage these two properties to deliver superior performance compared to other types of accumulators. Hence, they are by far the most commonly used type of accumulator. Moreover, they offer a good dynamic response. Remember, the design, manufacturing, and testing of accumulators should comply with applicable standards. Next, the typical design life for a hydraulic accumulator is 12 years.

An inert gas such as dry nitrogen should always be used as the gas medium. It may be noted that neither air nor oxygen should be used for charging an accumulator. If used, the gas medium may auto-ignite at high pressures, posing an explosion risk.

The gas-charged accumulators fall into two basic categories. They are: (1) Non-separator type and (2) Separator type.

Non-separator type Accumulator

Some of the earlier accumulators used in hydraulic systems were non-separator (or direct-contact) fluid-gas containers. Approximately half of the accumulators were filled with the fluid, and the other half with nitrogen gas, with no physical separation between them. The shut-off valve at the fluid port must be closed before stopping the system to prevent fluid and gas from escaping from the accumulator. In this type of accumulator, foaming and gas absorption into the fluid medium were commonly used. The gas used to get dissolved in the fluid and then carried away. Therefore, non-separator type accumulators are not used in modern hydraulic systems, but many such older units may still be in service.

Separator type Accumulator

A commonly used type of gas-charged accumulators is the separator type. In this type, there is a physical barrier, such as a diaphragm, a bladder (bag), or a floating piston, between the gas medium and the fluid medium it contains. This barrier exploits the gas's compressibility to absorb energy efficiently. Furthermore, a correctly sized port is provided for the hydraulic connection.

Pre-charging, Hydro-pneumatic Accumulators

A hydraulic accumulator must be pre-charged with clean nitrogen gas of at least Class 4.0, with a filtration level of 3 μm or better. A fill port is provided to supply nitrogen gas.

More details of pre-charging a hydro-pneumatic accumulator are given in Chapter 8.

Other Features, Hydro-pneumatic Accumulators

The installation, maintenance, and safety aspects of hydraulic accumulators are also very important for any hydraulic system that uses them and are discussed in the following sections and chapters.

Monitoring hydraulic accumulators via the industrial Internet of Things (IIoT) enables real-time, smart monitoring of pressure, temperature, and fluid levels. When sensors are integrated, they wirelessly transmit data to central systems, enabling quick detection of issues such as low pre-charge pressure or bladder breaches. This supports predictive maintenance and minimizes downtime, making operations smoother and more efficient.

Note: The Industrial Internet of Things (IIoT) involves connecting sensors and instruments to computers in industrial environments such as manufacturing. By gathering, analyzing, and responding to real-time data from machinery, IIoT helps boost efficiency and supports predictive maintenance.

Classification, Hydro-pneumatic Accumulators

According to the type of barrier used to separate the hydraulic fluid chamber and energy storage section, separator-type accumulators can be classified into:
- Piston Type
- Diaphragm type
- Bladder type
- Metal bellows type

The construction details of these types of accumulators are given in the following chapters. A comparison of these types of accumulators is presented in Table 7.1.

Chapter 3 | Piston Accumulators

A piston-type accumulator consists of a cylinder (or shell) with a finely finished internal surface, a freely moving metal piston with specialized seals, end caps, a fluid port, and a gas-charging valve. The piston separates the cylinder into two chambers. One chamber is to contain hydraulic fluid on the system side, and the other one is to hold nitrogen gas. The gas section is pre-charged with dry nitrogen gas. A schematic diagram of a piston accumulator is shown in Figure 3.1.

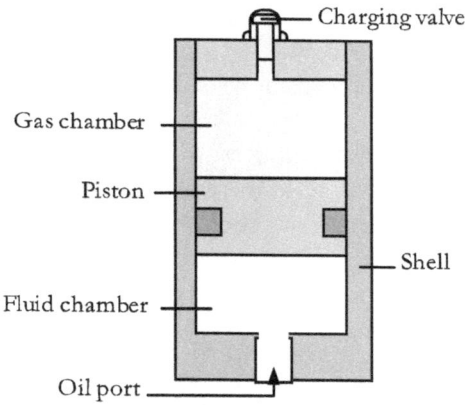

Figure 3.1 | A piston accumulator

Working of Piston Accumulators: The fluid chamber of the accumulator is connected to the hydraulic system, so that the accumulator draws fluid as system pressure increases, thereby compressing the gas and acting as an energy storage reservoir. When the system pressure drops, the pressurized gas expands, forcing the fluid back into the system. That is, the compressed gas provides a quick pneumatic spring action to force stored energy from the accumulator into the system.

Shell: The shell is manufactured from a piece of homogeneous, seamless tubing with a very finely machined internal surface. It is made of carbon steel, stainless steel, low-temperature carbon steel, or other materials. For use with a chemically aggressive fluid, the interior and exterior of the carbon steel shell can be optionally protected against corrosion with a plastic coating using epoxy or chemical plating using nickel or chrome. Alternatively, stainless steel can be used for accumulator parts that are liable to be exposed to a corrosive environment or subjected to high pressure.

Piston: The freely floating and movable piston with a pair of O-ring seals and guide rings separates the inner hollow space of the cylinder into two chambers. The piston must act as a leak-proof separation element. The piston is typically made from aluminum, carbon steel, or stainless steel. The piston must be lightweight for good acceleration. Piston diameters are typically available from 1.5" to 24" in a graded manner. Piston velocity is limited for various models from 1.5 to 16 ft/s (typical).

Sealing: The piston floats on low-friction O-ring seals to prevent metal-to-metal contact between the piston and the accumulator's internal surface. Seal materials for piston accumulators are formulated from the most advanced elastomers, capable of withstanding a wide temperature range from -40°F to 300°F. The advantages of using low-friction seals include no stick-slip, low wear, and high piston velocity (up to 16 ft/s).

Seal materials must also be compatible with a wide range of fluids, including mineral-based, fire-resistant, and synthetic fluids. Buna-N is the standard material used for piston seals and is suitable for most fluid power applications. Other materials, such as Viton or Polyurethane, can also be used as seal materials depending upon the system temperature extremes and the type of fluid used in the system.

End Caps: An accumulator is provided with end caps at both ends to close the cylinder. The end caps are typically made from carbon steel, stainless steel, or low-temperature carbon steel, with or without surface protection. An end cap contains an opening for the fluid port or gas port, as the case may be. Further, a piston accumulator with an internal diameter of up to 10" is fitted with a securing pin. This pin is intended to prevent the end cap from being removed incorrectly.

Fluid Port: The fluid chamber includes a standard fluid port with threads conforming to ISO (metric), DIN, ANSI (NPT), etc., or flanges conforming to DIN, ANSI, SAE, etc., for making a direct connection to the associated hydraulic system. The fluid port assembly is specifically designed to prevent turbulent flow, excessive pressure drop, and potential premature closure of the poppet valve.

Gas Valve: A piston accumulator can be recharged with nitrogen gas using a gas valve.

Sensors: The piston's position can be sensed or measured using a limit switch, proximity sensor, linear transducer, or a mechanical indicating rod.

Safety Devices: Many safety devices, such as burst discs, gas safety valves, and temperature fuse plugs, are available for use in piston accumulators. For example, the burst disc is designed to rupture and hence to keep the gas port open if the pressure on the gas side exceeds the maximum level. Further, a temperature fuse plug can be used to completely release the gas pressure when the temperature reaches an unacceptable level.

Installation of Piston Accumulators
Piston accumulators can be mounted in any orientation. It may be noted that an accumulator installed at an angle or in a horizontal position will become a trap for fluid contaminants. Therefore, vertical installation is preferable for a piston accumulator, with the fluid side at the bottom. Moreover, vertical installation is essential for a piston accumulator with a position indicator. Large piston accumulators with a piston diameter greater than 13.5" must only be installed vertically.

Clamping Supports
Mounting clamps with rubber inserts enable secure installation with minimal vibration. Clamping supports are preferably used near the end caps to mount a piston accumulator for proper support and isolation from system vibrations.

Back-up Gas Bottles
A backup gas bottle is simply a gas cylinder connected to the charging valve assembly of an accumulator. Backup gas bottles are used in combination with a piston accumulator to provide additional gas volume without increasing the piston size. This allows for large fluid delivery at a low cost with only a small pressure differential. The accumulator operates as before, but because of the greater gas volume, the expansion is greater, and therefore the fluid flows back into the system faster.

Advantages of Piston Accumulators:

Piston accumulators are compact devices. They can provide higher flow rates (typically up to 60 gpm) than comparable-sized gas-charged accumulators. For higher flow capacity, several accumulators or gas bottles can be connected in parallel. Other advantages of piston accumulators are their portability and ability to handle a broad range of temperatures. A piston accumulator has better damping capability than other types of hydraulic accumulators due to hydraulic leakage and friction between the piston and the shell.

Disadvantages of Piston Accumulators

The main disadvantages of piston accumulators are their susceptibility to fluid contamination and hysteresis caused by seal friction. They are also expensive to manufacture and have practical size limitations. Typically, a gas-charged piston accumulator costs twice as much as an equal-sized bladder accumulator. Further, it requires frequent pre-charging owing to increased gas leakage.

Moreover, it does not respond to transient pressures as quickly as an equivalent bladder accumulator does, due to the piston's greater mass. Also, remember that a piston accumulator cannot be used as a shock absorber because of the piston's inertia and the friction induced by piston seals.

Disassembly and Assembly Tips, Piston Accumulators

Repairing a high-pressure piston accumulator requires extreme caution because it stores energy even when the main system is shut down. Only authorized persons who are properly trained should disassemble or assemble a piston accumulator. Next, follow the instructions in the operating manual.

Before disassembling or assembling a piston accumulator, the system must always be depressurized. The fluid side should be relieved using a bleed valve, and the gas in the gas chamber should be depressurized and the charge valve opened before the accumulator is disassembled.

Ensure the piston moves freely before removing the end caps. A rod can be used to check the free movement or otherwise jamming of the piston.

The securing pin, if any, used to lock the end cap must be removed before removing the end cap.

Any welding, soldering, or mechanical work should not be carried out on the piston accumulator, on any account.

Compliance with Rules and Regulations

All responsible owners and operators of hydraulic equipment with accumulators, including piston accumulators, must adhere to the applicable rules and regulations governing pressure vessels at the location of the equipment's installation.

For example, the Pressure Equipment Directive (PED) in the EU and the ASME in the US.

General safety information for hydraulic accumulators is provided by relevant standards.

Some safety standards for high-pressure vessels are provided in Appendix 5.

Appendix 1 gives the specifications of piston accumulators.

Chapter 4 | Diaphragm Accumulators

A diaphragm (membrane) accumulator consists of two metallic hemispheres (shells), a flexible synthetic diaphragm, a fluid port, a gas valve, and mounting supports. The diaphragm is secured in between the two hemispheres. The diaphragm divides the accumulator's inner hollow space into a fluid chamber and a gas chamber. A schematic diagram of a typical diaphragm accumulator is shown in Figure 4.1.

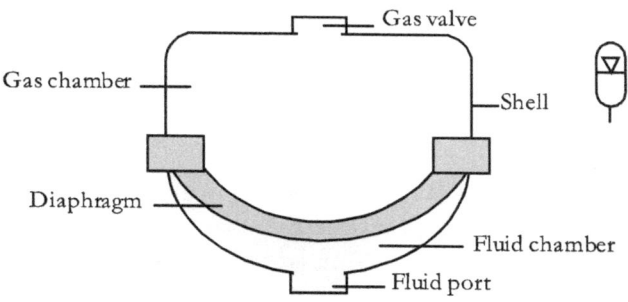

Figure 4.1 | A diaphragm-type accumulator

Working: When the system pressure increases, the fluid is drawn into the fluid chamber, and the gas gets compressed. Any pressure drop in the system causes the diaphragm to expand, forcing out the stored fluid from the accumulator back into the system. That is, the compressed gas provides a quick pneumatic spring action to force stored energy from the accumulator into the system.

Shells: The shells in an accumulator can be welded or screwed to form the accumulator housing. A lock nut can be provided in a screw-type accumulator to hold the upper and lower sections securely.

The shells are made of carbon steel, low-temperature steel, or high-tensile stainless steel. For use with a chemically aggressive fluid, the interior and, optionally, the exterior of the carbon steel shells can be provided with corrosion protection, a plastic coating using epoxy, or chemical plating using nickel or chrome. Alternatively, stainless steel can be used for the accumulator parts that are liable to be exposed to a corrosive environment or subjected to high pressure.

Diaphragm: The flexible diaphragm is a barrier that serves to divide the inner cavity of the accumulator into two chambers. One chamber is to contain hydraulic fluid on the system side, and the other one is to hold nitrogen gas. The barrier must separate the hydraulic fluid chamber and the gas chamber in a leak-free manner. In the screw-type accumulator, the diaphragm can be replaced. However, in a weld-type accumulator, the diaphragm cannot be replaced.

Diaphragm materials are developed from the most advanced elastomers capable of meeting a wide range of temperatures from -40°F to 350°F. They must also be compatible with a wide range of fluids, including mineral-based, fire-resistant, and synthetic fluids.

Buna-N is the standard material for bladders and is suitable for most fluid power applications. Other materials, such as Viton (Fluorine rubber, FKM), Butyl, or Hydrin, can also be used as the diaphragm material for an accumulator in a hydraulic system, depending on the system temperature limits and the type of fluid used.

If an accumulator discharges rapidly with a high ratio of the maximum system pressure P_2 to the pre-charge pressure P_0

(P_2/P_0), the gas tends to cool below the permitted temperature. This cooling may cause cracking of the diaphragm material.

Valve Poppet: A valve poppet is set into the base of the diaphragm to prevent extrusion of the diaphragm into the opening for the fluid passage, in case the diaphragm over-expands.

Fluid Port: The fluid chamber includes a standard fluid port with threads conforming to ISO (metric), DIN, ANSI (NPT), etc., for making a direct connection to the associated hydraulic system.

The fluid port assembly is specifically designed to prevent turbulent flow, excessive pressure drop, and potential premature closure of the poppet valve.

Gas Valve: In the weld-type and screw-type diaphragm accumulators, gas can be supplied or discharged via a normally closed gas charging valve with a knob. Alternatively, in a weld-type accumulator, the gas chamber can be fully charged with nitrogen gas, and the charging passage can then be completely sealed. The gas chamber must be pre-charged to a definite pressure.

Mounting: By design, a diaphragm accumulator can be mounted in any position. However, in a system where contamination is a severe problem, a vertical mount with the accumulator's fluid port oriented downward is usually preferred.

Support Clamps: Small accumulators with a nominal volume up to 0.5 gallons can be screwed directly inline.

An accumulator used in an application where strong vibrations are expected must be securely fastened with support clamps (fixing bands).

Key Features of Diaphragm Accumulators

Diaphragm accumulators are compact and lightweight. Therefore, they are fast-acting and exhibit no hysteresis.

They are not affected by contamination and exhibit consistent behavior under similar conditions.

They have high-pressure capabilities and are available with a pressure rating of up to 5000 psi.

They are essential components in hydraulic systems providing shock absorption, energy storage, and pressure stability features.

The speed of a diaphragm accumulator is governed by the gas, as there is no piston mass. Therefore, it reacts quickly to the changes in the system pressure. Hence, it is an excellent choice for an application requiring pressure pulsation damping.

Diaphragm accumulators are designed to handle small fluid volumes, typically less than one gallon.

Diaphragm accumulators have low maintenance requirements and a long service life.

These features make them indispensable for many applications.

Appendix 3 gives diaphragm accumulator specifications.

Chapter 5 | Bladder Accumulators

A bladder accumulator consists of a seamless cylindrical pressure vessel (shell), an internal elastomeric bladder (bag), a poppet valve, a fluid port, a charging valve, and clamps and brackets. The flexible elastomeric bladder separates the hydraulic fluid from the nitrogen gas. A schematic diagram of a bladder (or bag) accumulator is shown in Figure 5.1.

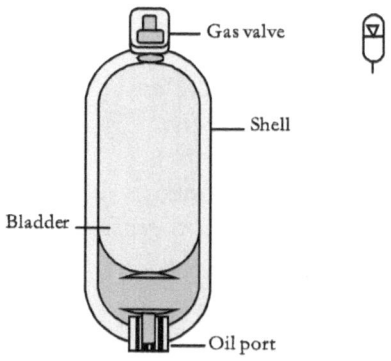

Figure 5.1 | A bladder accumulator

Working: When system pressure increases, the fluid is drawn into the fluid chamber, compressing the nitrogen gas. Any drop in system pressure causes nitrogen gas to expand, forcing the stored fluid from the accumulator back into the system.

Shell: The shell of a bladder accumulator is manufactured from homogeneous, seamless tubing that is usually heat-treated and stress-relieved, as required by relevant standards, to ensure excellent mechanical properties. Each end of the shell is formed in a hemispherical shape by spinning or forging.

The shell is made of carbon steel, high-tensile stainless steel, or low-temperature steel. For use with a chemically aggressive fluid, the interior and, optionally, the exterior of the carbon steel shell can be protected against corrosion, such as with a plastic coating with epoxy or chemical plating with nickel or chrome. Alternatively, stainless steel can be used for the accumulator parts that are liable to be exposed to a corrosive environment or subjected to high pressure.

Bladder: As stated earlier, the bladder divides the shell into two chambers, namely, the fluid chamber on the system side and the gas chamber inside the bladder. The fluid chamber is to contain hydraulic fluid on the system side. The gas chamber inside the bladder is pre-charged to a specific pressure. The bladder must separate the hydraulic fluid chamber and the gas chamber in a leak-free manner.

A full range of bladders is developed from the most advanced elastomers and can withstand temperatures from -50°F to 300°F. They must also be compatible with a wide variety of fluids.

Buna-N is the standard material for bladders and is suitable for most fluid power applications. Other materials, such as low-temperature Buna-N (ECO), chloroprene, nitrile, Viton (Fluorine rubber, FKM), Butyl, ethylene-propylene, and Hydrin, are also used as bladder materials.

If an accumulator discharges rapidly with a high ratio of the maximum system pressure P_2 to the pre-charge pressure P_0 (P_2/P_0), the gas tends to cool below the permitted temperature. This cooling may cause cracking of the bladder material.

Poppet Valve: The fluid chamber in a bladder accumulator is provided with a spring-loaded poppet valve. This valve closes when the pressure on the gas side exceeds that on the fluid side. In this way, the bladder is prevented from extruding into the downstream tubing if it overexpands. If the minimum operating pressure is reached, about 10% of the nominal volume should remain between the bladder and the poppet valve so that the bladder does not contact the valve during every expansion.

Fluid Port: The fluid chamber includes standard ports with threads conforming to ISO (metric), DIN, ANSI (NPT), etc., or special bolt-on flanges conforming to DIN, ANSI, SAE, etc., for making a direct connection to the associated hydraulic system. The fluid port assembly is specifically designed to prevent turbulent flow, excessive pressure drop, and potential premature closure of the poppet valve.

Bottom-repairable and Top-repairable Bladder Accumulator Models

Bladder accumulators are available in bottom-repairable and top-repairable configurations.

In a **bottom-repairable model**, the bladder can be inserted into the shell through the bottom opening, and the fluid port body and poppet valve assembly can be fixed to seal the accumulator. The bottom-repairable models are most popular.

A **top-repairable model** is designed as a cylindrical steel cylinder, an open-top bladder, an upper cap, a threaded ring assembly, and a fluid cap. The fluid cap with an opening is welded into the steel cylinder at the lower end. The open-top bladder can be inserted into the vessel and then can be retained and sealed by the upper cap and threaded ring assembly. A top-reparable model can be easily repaired without dismounting the accumulator. However, top-reparable models are comparatively expensive.

Clamps and Brackets: Clamps and brackets (console) can be used to mount accumulators for their proper support and isolation from system vibrations. They are zinc-plated to resist corrosion and can be easily bolted or welded to hydraulic systems.

A small accumulator can be fixed in place with a light-duty clamp, and a large accumulator can be mounted and supported with a heavy-duty clamp and a base bracket. Base brackets support large vertically mounted accumulators. The clamp is generally made of zinc-plated sheet steel or stainless steel strap. A rubber insert can be used in a clamp, and a rubber support ring can be incorporated into a bracket to absorb vibration and prevent noise transmission through metal-to-metal contact.

Charging Kits: The gas chamber is pre-charged with nitrogen gas to a certain pressure level using a nitrogen source and a charging kit. The charging kit consists of a normally closed charging valve with a knob, an adapter, and a pressure gauge. The adapter connects the charging valve to the gas port. An additional valve can also be provided to protect the pressure gauge.

Installation of Bladder Accumulator
Bladder accumulators can be installed vertically, horizontally, or at any angle, depending upon the application requirements. On a specific type of application, a particular position is preferable. For example, the vertical

position of the accumulator is preferable for energy storage applications, and any installation position from vertical to horizontal is appropriate for pulsation damping applications.

Advantages and Limitations of Bladder Accumulator

Bladder accumulators are fast-acting and do not exhibit hysteresis. Therefore, a bladder accumulator is an excellent option for an application requiring pressure pulsation damping and shock suppression.

Further, it is not susceptible to contamination and provides consistent behavior under similar conditions.

The main limitation of bladder accumulators is their larger size compared to other gas-charged types.

Features of Bladder Accumulators

Bladder accumulators are available as standard models or in custom-engineered designs.

Standard models are available in diameters from 2" to 24" with fluid capacities from 4 cubic inches to 300 gallons, and operating pressures up to 20000 psig.

Due to the limited volume capacity of a bladder accumulator, banks of bladder accumulators can be connected to a manifold to provide the desired quantity of fluid to a system. This arrangement of bladder accumulators can lead to physical space constraints in certain applications.

Appendix 2 gives the specifications of bladder accumulators.

Chapter 6 | Metal Bellows Accumulators

A metal bellows accumulator consists of a housing with metal bellows assembly. The metal bellows assembly divides the housing into a fluid section and a gas section. The bellows element is designed as either corrugated bellows or diaphragm bellows. The gas section is initially pre-charged and then permanently sealed. The working of a metal bellows accumulator is similar to that of piston accumulators. Figure 6.1 shows the schematic diagram of a metal bellows accumulator.

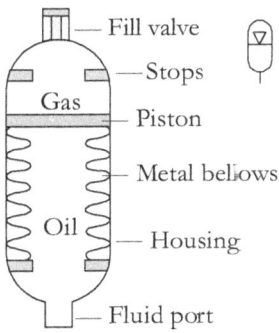

Figure 6.1 | A schematic of a metal bellows accumulator.

They are very reliable and usually, have a long service life. Moreover, they are gas-tight, maintenance-free, and media resistant over a broad range of temperatures. However, a metal bellows accumulator responds slowly to any pressure changes due to the increased mass of its piston and bellows.

They are used where a fast response time is not critical, yet reliability is essential. They are used in a wide range of applications, including aircraft systems, missile systems, ships, military ground vehicles, and chemical industries.

Chapter 7 | Comparison of Accumulators

Table 7.1 gives a comparison of diaphragm, bladder and piston type hydraulic accumulators. The values given are typical.

Table 7.1 | Comparison of accumulators

Parameter / property	Diaphragm	Bladder	Piston
Size	≤0.92 gallons	≤14 gallons	≤350 gallons
Working pressure	3500 psi	10000 psi	36000 psi
Flow rate	≤40 gpm	≤240 gpm	≤240 gpm
Compression ratio	8:1	4:1	10:1
Flow rate (Typical)	≤30 in³/s	≤4 gallons/s	≤57 gallons/s
Seal materials	Buna-N Viton Butyl Hydrin	Buna-N Butyl Viton Hydrin	Buna-N Viton Polyurethane
Temperature limits	-40°F to 350°F	-50°F to 300°F	-40°F to 300°F
Heaviness	Light-weight	Medium-weight	Heavy
Cost	Low	Medium	High
Application	Suitable for small volume and flow rates	Best for general purpose applications	Best for large volumes or high flow rates
Shock suppression	Good	Good	Not good
Mounting position	Any position	Any position	Only vertical
Fluid port connection	SAE, NPT	SAE, NPT	SAE, NPT

Chapter 8 | Pre-charging of Accumulators

A hydro-pneumatic accumulator must be filled with dry inert gas, such as nitrogen, at the specified pre-charge pressure while there is no fluid in the hydraulic part. Before initial start-up or whenever the pre-charge pressure falls below the specified value due to gas leakage, it must be set to meet the application requirements or as specified by the machine manufacturer.

The pre-charge pressure is an essential parameter for the gas accumulator, as it, along with the accumulator volume, determines the maximum amount of hydraulic energy that can be stored. A very low pre-charge pressure leads to insufficient energy storage, poor shock absorption, and/or damage to parts. On the other hand, a high pre-charge pressure can damage accumulator parts and reduce its service life.

Pressure Levels in a System with Accumulators
A hydraulic system with an accumulator involves three pressure levels. These pressure levels are explained below.

Maximum System Pressure (P_2): The maximum system pressure is set by the pressure relief valve in a hydraulic system with accumulators.

Minimum System Pressure (P_1): This is the lowest operational pressure required to operate the associated load effectively. It must be maintained at a level high enough to enable the accumulator to fulfill its designated function. Additionally, it prevents the moving element in a hydro-pneumatic accumulator from reaching the bottom level. This pressure is usually a design parameter.

Pre-charge Pressure (P_0): The initial pressure of nitrogen gas introduced into a hydraulic accumulator before hydraulic fluid enters the system. If the pre-charge pressure, especially in a hydro-pneumatic accumulator, exceeds its limits, its elastomeric parts are liable to damage.

Pre-charge Pressure Setting

The pre-charge pressure depends on the type of accumulator and application and is recommended by the manufacturer.

Energy Storage Applications: For energy storage applications, the pre-charge pressure P_0 can typically be 80 to 90 percent of the system's minimum working pressure P_1.

Shock Absorption Applications: The pre-charge pressure P_0 for a shock absorber or a pulsation compensator can be 65 to 80 percent of the minimum operating pressure P_1.

Additional Limitation: The gas pre-charge pressure is further limited by the maximum pressure ratio $P_2:P_0$, which should not exceed 4:1. [$P_2/P_0 \leq 4/1$]

That is, $P_2:P_0 \leq 4:1$

Summary: The summary of the pre-charge pressure level calculation is given in Table 8.1

Table 8.1 | Pre-charge pressure calculation

Function	Pre-charge pressure (P_0)
Energy storage	80 to 90% of P_1
Shock absorption or pulsation dampening	65 to 80% of P_1

Pre-charge During Accumulator Transportation

Accumulators should be pre-charged with nitrogen gas at a safe level, low enough to comply with regulations during transit. Diaphragm accumulators are generally delivered without pre-charge pressure. Bladder accumulators are generally delivered with a nitrogen pre-charge pressure of approximately 30-70 psi. This minimal pressure protects the bladder from damage.

Accumulator Pre-charging Procedure

A gas-charged accumulator is delivered with low gas pressure. It can be charged before or after being installed on the system. Before charging, it is advisable to pour some fluid into the accumulator to allow the fluid to coat the inside of its shell. This fluid layer provides the initial lubrication between the bladder and the shell.

Initially, a hydro-pneumatic accumulator should be pre-charged with clean, dry nitrogen gas of class 4.0 purity (N2 content 99.99% by volume) using a charging kit. If the nitrogen gas is allowed to flow too rapidly into the accumulator, it can lead to the chilling of the polymeric material of the accumulator's diaphragm or bladder. This chilling effect may result in the immediate brittle failure of the polymeric material.

Figure 11.1 shows the schematic diagram indicating the arrangement for pre-charging the accumulator. This package mainly consists of a nitrogen bottle, a charging manifold, a gas chuck, and connecting hoses. The manifold includes a bleed valve and a pressure gauge. The nitrogen bottle has a gas valve and a pressure regulator. The accumulator is usually provided with a gas valve and a protective cap.

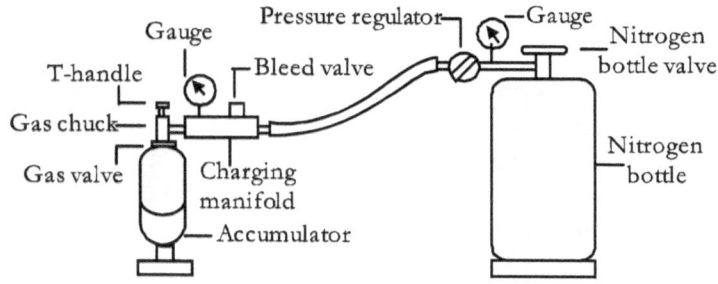

Figure 11.1 | A setup for pre-charging an accumulator

The pre-charging procedure is listed below:
-Attach the charging kit hose to the gas chuck on one side and the regulator on the other side
-Close the bleed valve on the charging manifold
-Attach the chuck to the accumulator gas valve
-Open the gas valve by turning the chuck's T-handle
-Open the nitrogen bottle valve slowly and fill the accumulator to the desired pre-charge pressure
-Close the nitrogen bottle valve
-If the desired pre-charge is exceeded, open the bleed valve to relieve the excess pressure
-Close the bleed valve and the gas valve
-Remove the gas chuck

Checking of Pre-charge Pressure
The pre-charge pressure of an accumulator should be checked regularly. Typically, the first check should be carried out after one week of operation, the second after three months, the third after one year, and thereafter continue once every year. If the pre-charge is low, investigate the cause and rectify it. Possible causes of the low pre-charge pressure include gas leakage, a damaged gas valve, or a damaged bladder/diaphragm.

Chapter 9 | Safety Requirements of Accumulators

A hydraulic accumulator is a pressure vessel that stores a large amount of potential energy for subsequent release to perform useful hydraulic functions. Accumulators can be dangerous to personnel and property if they discharge the stored pressure inadvertently. Moreover, they are subject to regulations applicable to the place of their installation.

Therefore, it is necessary to isolate the accumulators from the associated systems and discharge their trapped pressure during maintenance or emergency periods.

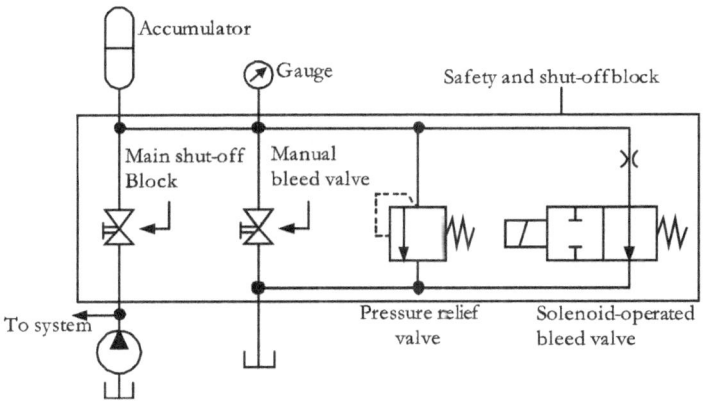

Figure 9.1 | A circuit layout of a safety-and-shut-off block for a hydraulic accumulator

Typically, safety devices must be incorporated into an accumulator to provide shut-off, pressure limiting, and pressure relief. It is recommended to always use a safety-and-shut-off block along with an accumulator to protect the system and personnel from hazardous stored energy.

Figure 9.1 shows the circuit diagram of a safety-and-shut-off block.

Safety-and-Shut-off Block

Figure 9.2 shows a schematic diagram of a safety-and-shut-off block connected to an accumulator to protect it against pressure peaks. It is a multifunctional valve system connected to a hydraulic accumulator.

Further, it consists of a shut-off valve, a manual bleed valve, a pressure relief valve, an optional 2-way solenoid-operated bleed valve, and a pressure gauge. It can be zinc-plated for optimum corrosion resistance.

It is possible to configure the block in a modular fashion, with a host of connection options, enhancing its versatility for mounting.

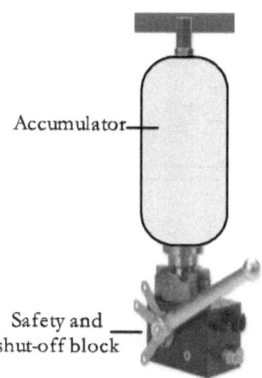

Figure 9.2 | A schematic of a safety-and-shut-off block

Shut-off Valve: The shut-off valve instantly isolates the accumulator from the hydraulic system for maintenance purposes or emergency use.

Bleed Valves: They are available with manual and electrical control options for the safe discharge of pressurized fluid into the reservoir.

Manual Bleed Valve: Once isolated, the accumulator can be safely discharged to the reservoir through the manual bleed valve.

Solenoid-operated Solenoid Valve: If used, the optional solenoid-operated bleed valve automatically releases the accumulator's stored energy during an emergency shutdown or loss of electrical power.

Pressure-relief Valve: A pressure-relief valve protects the accumulator from over-pressurization.

Pressure Gauge Port: An accumulator safety block includes a port for a pressure gauge to monitor system pressure.

Other Important Features / Requirements of Safety Blocks

The fluid port threads can be as per SAE, BSPP, etc.

The safety and shut-off block should always be mounted close to the accumulator.

Remember, the commissioning and maintenance of the safety block and associated equipment must be performed only by qualified technical staff.

The specification parameters for accumulators and safety blocks are given in Appendix 4.

Chapter 10 | Applications of Accumulators

Accumulators are widely used in industrial and mobile applications for shock suppression, energy storage, leakage compensation, and energy recovery.

Accumulators are found on industrial machinery and mobile equipment, as well as in marine, oil and gas, and aerospace applications.

For example, they are used in industrial applications such as machine tools, steel production, metal-forming machinery, foundries, paper production, power transmission, injection molding, and die casting.

They are used in mobile applications such as mining, construction, forestry, and agriculture.

For shock suppression, they are used in large hydraulic presses, vehicles, construction equipment, offshore equipment, and mining equipment.

An accumulator can supplement the pump flow in a hydraulic application, such as an aircraft landing gear, which requires a considerable volume of fluid.

The use of accumulators as an energy storage device is convenient for completing a working cycle in the event of failure of the primary power source.

The inclusion of accumulators in applications enables the use of small pumping stations, such as cranes, dock gates, lifts, and forging presses, that do not use the entire pump flow for most of the machine cycle time.

The weight-reduced diaphragm, bladder, and piston accumulators are used in the aircraft, offshore, city bus, garbage truck, wind power, automotive, and railway vehicle industries for energy storage and to reduce energy consumption.

Summary of Accumulator Applications

A summary of end-user applications of hydraulic accumulators is given in Table 11.1.

Table 11.1 | End-user applications of accumulators

Accumulator function	End-user applications
Shock Absorption and Pulsation Dampening (Smooth out hydraulic actuator movement)	in construction, offshore, and mining equipment, presses, and vehicles
Energy Savings and Auxiliary Power (For supplying high-pressure energy for quick, short-stroke cycles)	Plastic injection molding, presses, forming machines, mining, excavators & loaders, and aircraft landing gear
Leakage Compensation (For systems requiring high pressure over long, continuous periods)	Manufacturing assembly lines (Robotic welding and assembly setups) and testing equipment (Maintaining high pressure for long-duration leak tests)
Emergency and Safety Systems (For backup power)	Aviation (Landing gear), Oil & Gas (Blowout Preventer - BOP), and industrial machinery (Emergency brakes, safety door closing system)

Chapter 11 | Basic Accumulator Circuits

Hydraulic accumulators perform many functions in hydraulic systems. They are basically employed for shock absorption and energy storage. The following sections explain many typical hydraulic circuits for realising these functions.

An accumulator as a Hydraulic Shock Absorber

Figure 11.1(a) shows a simple hydraulic circuit used for the direction control of a double-acting cylinder using a 4/3–way, closed-centre valve.

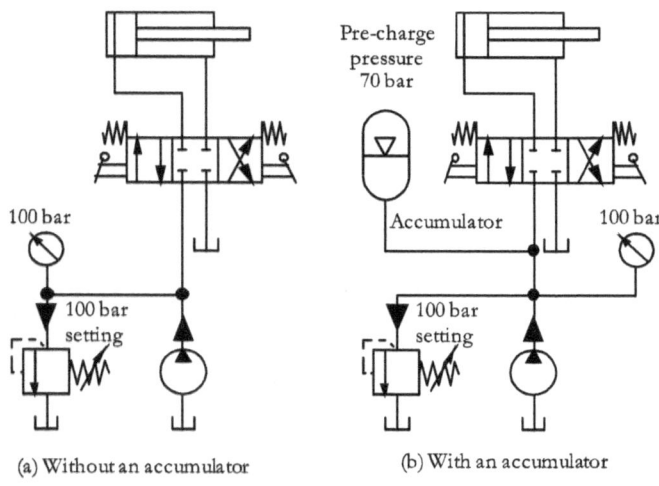

(a) Without an accumulator

(b) With an accumulator

Figure 11.1 | Basic hydraulic circuits

As we are aware, momentary high-pressure surges or shock waves are likely to be generated in the system when the valve is rapidly shifted. These shock pressures result from the failure of the associated PRV to drain the high-pressure fluid from the circuit quickly enough. These high-pressure surges can be dangerous to personnel and equipment.

A correctly-sized accumulator can be connected to the circuit, as shown in Figure 11.1(b), to suppress these shock pressures. The accumulator absorbs the shock pressures whenever they appear in the circuit. However, it can be observed that this circuit is unable to store the excess energy added to the accumulator, as the pressure relief valve slowly drains the excess fluid back into the system reservoir.

Other functions, such as storing energy, unloading the pump upon reaching the system pressure, and setting the discharge flow rate from the accumulator to suit the system requirement, can be achieved by using additional components. The following sections explain the circuits for realizing these accumulator functions.

An accumulator as an Auxiliary Power Source

Figure 11.2 gives different positions of a hydraulic circuit with a fixed-displacement pump, reservoir, unloading valve, and accumulator. The accumulator should be sized appropriately to store energy during periods of low demand and to provide auxiliary power during periods of high demand.

Figure 11.2(a) shows the position of the circuit when the 4/3-way, closed-center valve is pulled to its neutral position. In this position, the pump charges the accumulator and builds pressure through the check valve. The unloading valve remains closed until the set pressure is reached. The unloading valve opens and returns the pump flow to the reservoir at low pressure when the valve's pressure setting is reached. Unloading the pump flow allows the pump to operate at a minimal load. The isolation check valve prevents the accumulator fluid from flowing back to the pump and traps the pressurized fluid in the accumulator.

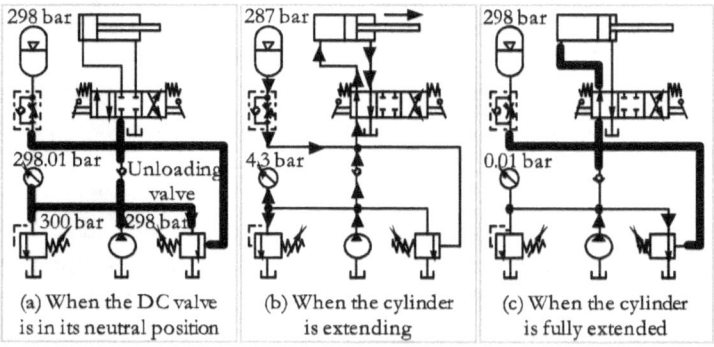

| (a) When the DC valve is in its neutral position | (b) When the cylinder is extending | (c) When the cylinder is fully extended |

Figure 11.2 | Three positions of a circuit with an accumulator acting as an auxiliary power source

Figure 11.2(b) shows the position when the 4/3-way valve is shifted to its left envelope for the forward stroke of the cylinder. During this position, the pump flow is supplemented by the accumulator, which releases the stored energy. Remember that the accumulator contains a volume of fluid under elevated pressure that can discharge almost instantaneously into the system. Therefore, when the accumulator is releasing stored energy, a throttle valve is required to set the flow rate within the system's requirements.

Figure 11.2(c) shows the position of the circuit when the cylinder is fully extended and the accumulator is fully charged. When the set pressure is reached, the unloading valve opens, unloading the pump to the reservoir at low pressure.

When the 4/3-way DC valve is shifted to its right envelope during the cylinder's return stroke, similar actions, as explained in the previous paragraphs, can be expected from the accumulator and the unloading valve.

Accumulator Circuit with an Unloading and a Dump Valve

It is necessary to discharge the pressurized fluid from the accumulator, either manually or automatically, when the system pump stops, as a safety precaution. A solenoid- or pilot-operated dump valve can automatically discharge the stored fluid when the pump stops. The dump valve is designed to close when the pump runs and open when the pump stops.

Parts (a) and (b) of Figure 11.3 show different positions of the hydraulic circuit with a fixed-volume pump, reservoir, accumulator, check valve, unloading valve, and 2/2-way pilot-operated dump valve.

Figure 11.3(a) shows the position of the circuit as soon as the pump is switched on. The flow is directed to the accumulator through the opened check valve. The dump valve closes immediately and remains closed even as the accumulator continues to charge.

When the set pressure of the unloading valve is reached, the valve opens and unloads the pump flow to the reservoir at low pressure. It may be noted that the dump valve remains closed, even when the unloading valve is relieving the excess pressure.

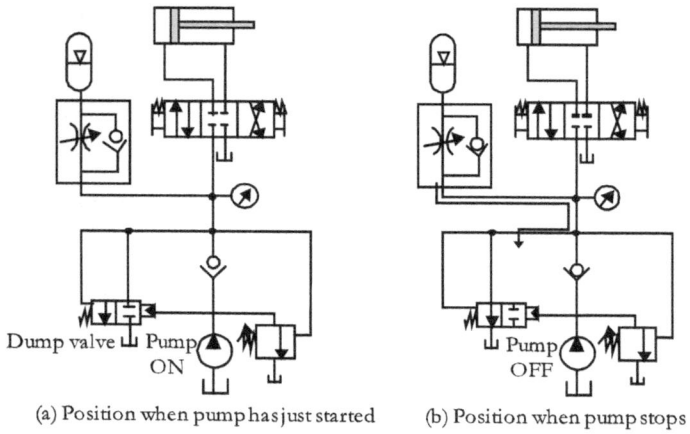

(a) Position when pump has just started   (b) Position when pump stops

Figure 11.3 | Two positions of an accumulator circuit with an automatic pilot-operated dump valve

Figure 11.3(b) shows the circuit's position immediately after the pump is switched off. Here, the pilot pressure to the dump valve drops, allowing its spool to shift position. The dump valve now opens, allowing pressurized fluid in the accumulator to flow directly to the reservoir. Therefore, the accumulator can discharge quickly and automatically when the pump is turned off, making it safe to perform maintenance on the system.

A pilot-operated dump valve works well in most cases, but it can cause problems in some situations. For example, if the dump valve fails to open when the pump stops, the circuit is unsafe. This possibility is a safety hazard to maintenance personnel. Therefore, it is advisable to be cautious and check an accumulator circuit for the trapped pressure before working on it, even if an automatic unloading arrangement is provided.

Chapter 12 | Maintenance of Accumulators

Following safe practices when working with accumulators and properly maintaining them are the most critical activities to ensure the safety of the associated equipment and personnel.

Maintenance personnel should be familiar with the rules governing pressure vessels, such as accumulators.

It is also essential to take all safety precautions in hydraulic accumulators to prevent hazardous stored energy, including built-in PRVs, shut-off valves, and solenoid- and/or manually operated bleed valves.

Always use proper PPE during accumulator maintenance.

Never use oxygen or compressed air in accumulators as this can be explosive. Only use dry nitrogen for pre-charging.

Therefore, an accumulator used in a hydraulic system should be designed to shut off the accumulator, discharge trapped fluid, and protect the system in an emergency or during a system shutdown.

The solenoid-operated bleed valve in the accumulator automatically releases the energy trapped in the accumulator during an emergency or during shutdown of the associated system.

The following sections present general maintenance guidelines for hydraulic accumulators, details of their installation, and pre-charging of gas-loaded accumulators.

Guidelines for the Maintenance of Accumulators

The following bulleted lines list some necessary maintenance and safety guidelines for the hydraulic circuit with an accumulator. These guidelines are the most general and are not intended for any specific hydraulic machine.

- Only qualified maintenance technicians must carry out the maintenance work on the system with the accumulator.

- Attach a warning sign, such as 'ATTENTION: System with Accumulator', close to the accumulator.

- The accumulator should be provided with the safety valve block with a PRV, shut-off valve, and bleed valve.

- Always depressurize and isolate the accumulator before servicing the system.

- Never start the system before charging the accumulator.

- Always maintain the maximum working pressure, pre-charge pressure, and operating temperature of the accumulator within acceptable limits.

- If the accumulator is a gas-charged one, make sure that the charging and discharging rates of the accumulator are restricted to reasonable values to avoid damage to the accumulator and other system components.

- Ensure that the accumulator is manufactured, tested, and certified as per the statutory standards, like ASME.

- Do not operate a system with an accumulator circuit until the accumulator is securely anchored to a stable structure.

Accumulator Installation

An accumulator in the hydraulic circuit should be installed as close as possible to the source of shock or potential energy to minimize pressure loss along the line between them.

It must be correctly installed in an easily accessible place using robust collars.

The markings engraved on the accumulator must remain visible after its installation.

Typically, an accumulator is installed vertically with the fluid connection port at the bottom.

However, if the accumulator must be mounted horizontally due to space constraints, efficiency will be reduced and service life shortened. Further, the horizontal arrangement could trap contaminants.

Inspection of Accumulators

Periodic inspection, especially of a large-volume or high-pressure accumulator, is required to detect any developing symptoms of potential malfunction.

The inspection may be carried out at predetermined intervals, such as annually or as specified by the manufacturer.

Remember, the owner and the operator of the equipment are responsible for ensuring the safe operation, thoroughness, and frequency of the inspection of the pressure vessel.

Chapter 13 | Accumulator Sizing

Gas-charged accumulators are widely used across many industries to store energy for intermittent duty cycles or to provide standby power. They are typically rated by the gas volume they can hold when all the fluid has been discharged. A hydro-pneumatic accumulator must be sized optimally and pre-charged to the correct pressure to meet the assigned function.

The accumulator sizing is affected by the gas's compression (or expansion) process. If charging (or discharging) of the accumulator occurs slowly, there is sufficient time for the accumulator wall to add (or remove) heat to maintain a constant gas temperature (an isothermal process). If charging (or discharging) occurs rapidly, there is insufficient time for sufficient heat transfer through the accumulator walls (an adiabatic process). These conditions are theoretical, and the actual charging (or discharging) process occurs between them. The actual charging (or discharging) process is called the polytropic process.

It is, however, possible to state with reasonable accuracy that when an accumulator is used as a volume or leakage compensator, the condition is isothermal. On the other hand, when an accumulator is used in other applications, such as energy storage, a pulsation damper, an emergency power source, or a shock absorber, the condition is close to adiabatic.

Calculating the accumulator's capacity for the system requires intimate knowledge of the system's operating conditions. Then, using any one of the mathematical models described below, the size of the accumulator can be calculated. Manufacturers also offer sophisticated software packages for accurately determining an accumulator's capacity. Specific definitions would be useful for further explanation:

Maximum System Pressure (P_2): This is the maximum (nominal) pressure developed in the system, as per the setting of the system relief valve. It is usually the no-flow rating of the hydraulic pump.

Minimum System Pressure (P_1): This is the minimum pressure the accumulator must maintain in the hydraulic system. This pressure is a design requirement used to size the accumulator.

Pre-charge Pressure (P_0): It is the pressure of the nitrogen in an accumulator without any hydraulic fluid in the accumulator. The pre-charge pressure determines the amount of fluid that an accumulator can hold at the system pressure and the desired minimum hydraulic system pressure.

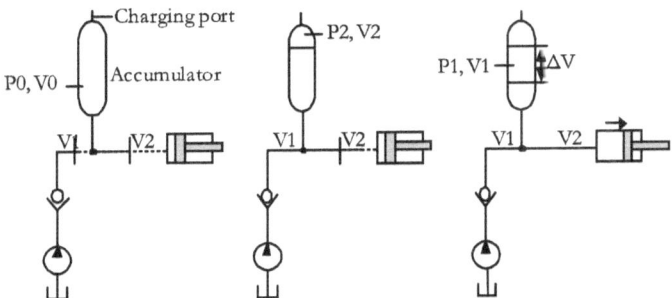

(a) Pre-charge position (b) Fully-charged position (c) Minimum charge position

Figure 13.1 | A schematic diagram showing three positions of a gas-charged accumulator connected in a hydraulic system

Figure 13.1 shows different positions of a hydraulic system with a pump, cylinder, check valve, and gas-charged accumulator for supplementing the power. Directional control (DC) valves V_1 and V_2 are also used in the circuit. Let D be the bore diameter, and S be the stroke length of the cylinder. Figure 13.1(a) shows the position of the accumulator when it is almost pre-charged. Let P_0 [= (0.95 to 0.97) x P_1 (i.e., Minimum system pressure)] be the pre-charge pressure in the accumulator, and V_0 be its corresponding volume. V_0 is the required accumulator size.

Figure 13.1(b) shows the hydraulic circuit configuration when the pump is on, with the DC valve V_1 open and V_2 closed. During this period, the accumulator is charged to the maximum pressure set by the pressure relief valve (not shown) in the system. Let P_2 be the maximum charge pressure in the accumulator, and V_2 be the corresponding volume during this stage.

Figure 13.1(c) shows the position of the circuit when both the DC valves V_1 and V_2 are open, and the cylinder has just reached the end of its stroke. During this period, the accumulator discharges fluid into the system, reducing its pressure. Let P_1 be the pressure in the accumulator during this stage, and V_1 be the corresponding volume. It may be noted that the fluid capacity available in the accumulator to drive the cylinder is $(V_1 - V_2)$.

In the isothermal process, the compression and expansion of the gas in an accumulator occur slowly, so that an approximately constant temperature condition can be assumed, as complete heat transfer between the gas and the environment is possible. In an adiabatic process, the compression and expansion of the gas are rapid, so no heat transfer occurs between the gas and its surroundings. The required accumulator size for various charging and discharging conditions can be derived in the following ways.

Charging and Discharging under Isothermal Conditions

The perfect gas laws govern the compression and decompression of the nitrogen gas contained in the accumulator. Using the Boyle-Mariotte's law for ideal gases, assuming slow charging and slow discharging, to allow the gas in the accumulator to maintain its temperature close to a constant, we get the size of the accumulator V_0 from the following equations:

$$P_0 V_0 = P_1 V_1 = P_2 V_2$$

That is, $V_1 - V_2 = [(P_0 V_0) / P_1] - [(P_0 V_0) / P_2]$

$$= V_0 [(P_0 / P_1) - (P_0 / P_2)]$$

Therefore,

Accumulator volume, V_0
$$= [V_1 - V_2] / [(P_0 / P_1) - (P_0 / P_2)]$$

$(V_1 - V_2)$ is the fluid volume to be supplied by the accumulator for the extension stroke ($x\%$ of the full cylinder volume). That is,

$$V_1 - V_2 = (x \%) \times (\prod D^2 / 4) \times S$$

As the gas is pressurized, its temperature is likely to rise, and the volume of fluid entering the accumulator is lower than the calculated amount. This limitation can be compensated for by increasing the accumulator capacity by about 5%. If there are many actuators in the system, consider the peak loading when sizing the accumulator.

Charging and Discharging under Adiabatic Conditions

Most fluid power designers use the ideal gas laws for the design calculations of accumulators. However, the primary gas laws do not apply when there is little or no heat transfer into or out of an accumulator, as in a shock absorber, pulsation damper, or emergency power source. Remember, today's hydraulic systems move faster with higher cycle rates. There is a short time for heat to enter or leave the accumulator, so we assume that the gas compression and expansion are adiabatic – that is, no heat is transferred into or out of the accumulator. Assuming quick charging and quick discharging, we get the size of the accumulator V_0 from the following equation:

$$P_0 V_0^n = P_1 V_1^n = P_2 V_2^n$$

Where n is the polytropic exponent (n=1.4 for diatomic gas)

The P-V diagram in Figure 13.2 shows the pressure-volume relationship in a gas-charged accumulator.

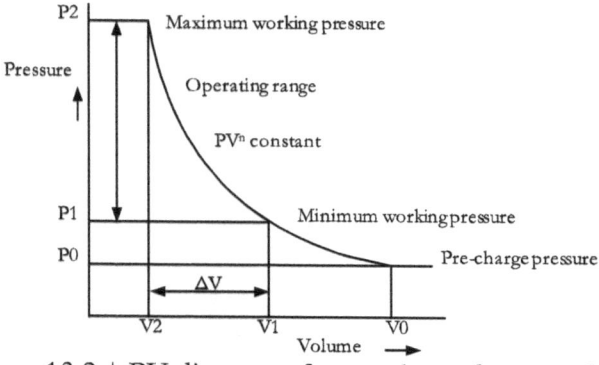

Figure 13.2 | PV diagram of a gas-charged accumulator

Similar to what is shown for finding the size of the accumulator with isothermal charging and discharging process, we have the size of the accumulator with charging and discharging under the adiabatic condition as:

Accumulator volume, V_0, adiabatic
$= [V_1 - V_2] / [(P_0 /P_1)^{1/n} - (P_0 /P_2)^{1/n}]$

Note:
- Use the ratio of absolute pressures, not the gauge values.

- Both the operating temperatures and pressures influence the accumulator calculations.

Slow Charging and Quick Discharging
Accumulator capacity when the charging process is slow (isothermal), and the discharging process is quick (adiabatic), can be calculated from the following equation:

Accumulator volume, V_0
$= [V_1 - V_2] / \{(P_0 /P_2)^{1/n} - [(P_2 /P_1)^{1/n} - 1]\}$

Temperature Influence, Accumulator Sizing
The operating temperature is liable to change significantly during the charging and discharging cycles of an accumulator. This temperature variation should be accounted for when calculating the accumulator volume. Let the capacity of the accumulator be V_0 at the temperature T_1 (K), and the capacity of the accumulator has increased to V_{oT} when the temperature has risen to T_2 (K). The capacity (V_{oT}) can be calculated by using the following formula:

$$\text{Accumulator volume, } V_{oT} = V_0 \times (T_2/T_1)$$

Correction Coefficient at Higher Pressures

The nitrogen gas in accumulators does not behave according to the ideal gas laws at pressures above 2900 psi. The capacity of the accumulator (V_{oP}), at higher pressures, under isothermal and adiabatic conditions, can be calculated by using the following formula:

Accumulator volume, $V_{oP} = V_0 / C_i$ (isothermal condition)

$= V_0 / C_a$ (adiabatic condition)

Where C_i is the isothermal correction coefficient, and C_a is the adiabatic correction coefficient, both must be determined from the manufacturer's standard charts.

Useful Tips, Sizing of Hydro-pneumatic Accumulators

- To achieve the best utilization of the accumulator volume possible as well as long service life, the pre-charge pressure at the maximum operating temperature should be about 90% of the maximum system pressure.

- The maximum system pressure (P_2) is not to exceed four times the pre-charge pressure (P_0) [$P_2 \leq 4 \times P_0$] to prevent excess strain on the elasticity of the bladder.

- The smaller the difference between P_1 and P_2, the longer the service life of the bladder. However, the degree of utilization of the maximum storage capacity will be reduced.

Example 13.1 | A molding press must remain closed during a 90-minute curing phase at a constant pressure of 2756 psi. After the mold has been closed, the pump is switched off. The expected oil leakage rate is 0.244 in^3/min, and the minimum permissible pressure during the curing period is 2727 psi. What is the capacity of a gas-charged accumulator used to compensate for the leakage? Assume the pre-charge pressure as 2582 psi.

Solution
Leakage rate = 0.244 in^3/min
Curing period = 90 min

Working pressure, P_2 =2756 psi = 2770.5 psi (absolute)

Minimum permissible pressure, P_1
=2727 psi = 2741.5 psi(a)

Pre-charge pressure, P_0 = 2582 psi = 2596.5 psi (absolute)

Let V_0, V_1, and V_2 be the accumulator volumes at their pre-charge, fully charged, and minimum charge positions, respectively.

Total leakage during the curing period=0.244x90 =21.96 in^3

Therefore, $V_1 - V_2$ = 21.96 in^3

Capacity of the accumulator, V0
= (21.96)/ [(2596.5/2741.5) – (2596.5/2770 5)]
=2215 in^3 = 9.6 gallons
A standard capacity accumulator closest to the calculated value must be selected.

Example 13.2 | A gas-charged accumulator used as an emergency power source charges and discharges quickly at the rate of 122 in^3 in 2 seconds. It has a minimum working pressure of 1015 psi and a maximum working pressure of 1958 psi, and an operating temperature range of 20°C to 80 °C. Assume the pre-charge pressure as 95% of the minimum pressure. Calculate the accumulator capacity.

Solution
Charging/discharging rate =122 in^3 in 2 seconds
Minimum working pressure, P_1= 1015 psi = 1029.5 psi (a)
Minimum working pressure, P_2 =1958 psi = 1972.5 psi (a)
Minimum temperature, T_1 = 20°C = 20+273 K =293K
Maximum temperature, T_2 = 80°C = 80+273 K =353K
Assume the value of the polytropic exponent as 1.4.
Pre-charge pressure, P_0 =0.95x1958 psi = 1860 psi
= 1874.5 psi (absolute)

As the accumulator is used as an emergency power source, its charging and discharging are quick, and the charging and discharging operations can be considered adiabatic. Therefore,

Accumulator capacity, adiabatic, V_0
=122/ {(1874.5 /1029.5)1/1.4 – (1874.5 /1972.5)1/1.4}
=214 in^3 =0.93 gallon

Considering the temperature variations,

Accumulator capacity, V_{oT} = V_0 x (T_2/T_1)=214 x (353/293)
=257.8 in^3 = 1.11 gallons
A standard capacity accumulator closest to the calculated value must be selected.

Example 13.3 | A gas-charged accumulator must discharge 281 in^3 of fluid for 3 seconds with a change of pressure from 4061 psi to 3017 psi. The loading time is 4 minutes. Define the capacity, keeping in mind that the ambient temperature changes from 20°C to 50°C. Assume the adiabatic and isothermal correction factors for higher pressures: Ca = 0.72 and Ci = 0.83, respectively.

Solution
Discharge rate = 281 in^3 in 3 s
Minimum operating pressure, P_1=3017 psi = 3031.5 psi(a)
Maximum operating pressure, P2=4061 psi = 4075.5 psi(a)
Minimum temperature, T_1 = 20°C = 273+20 K = 293 K
Maximum temperature, T_2 = 50°C = 273+50 K = 323 K
Correction coefficient, adiabatic, Ca = 0.72
Correction coefficient, isothermal, Ci = 0.83
Assume the value of the polytropic exponent as 1.4.
Pre-charge pressure, P_0 =0.95x3017 = 2866 psi
= 2880.5 psi (absolute)
Correction coefficient, C_{av} = (0.72 + 0.82)/2 = 0.77

As the accumulator storage is slow (isothermal), and discharge is quick (adiabatic), accumulator capacity can be calculated from the following equation:

Accumulator capacity, V_0
= 281 / {(2880.5/4075.5)1/1.4 x [(4075.5/3031.5)1/1.4-1]}
= 1566 in^3 = 6.7 gallons

Now, considering the correction coefficient for high pressure and the temperature change, we have:

Accumulator capacity, V_{oT} = (V_0/C_{av})x(T_2/T_1)

$= (1566/0.77)\text{x}(323/293) = 2242\,\text{in}^3 = 9.7\,\text{gallons}$

An accumulator closest to the calculated value must be selected.

14 | Objective Type Questions

1. A function of hydraulic accumulators is:
a) Limiting pressure
b) Dampening pulsations
c) Controlling speed
d) Conditioning fluid

2. Mark, the incorrect statement about hydraulic accumulators:
a) Piston accumulators have a higher compression ratio.
b) Piston accumulators can be used as a pulsation damper or a shock absorber.
c) A diaphragm accumulator is an excellent choice for applications requiring pressure pulsation damping.
d) The diaphragm accumulators are fast-acting and do not exhibit hysteresis.

3. The best type of hydraulic accumulator for large volumes or high flow rates is:
a) Diaphragm
b) Bladder
c) Piston
d) Spring-loaded

4. The gas used in a hydro-pneumatic accumulator is:
a) Air
b) Nitrogen
c) Oxygen
d) All of the above

5. Proper maintenance of gas-charged accumulators involves:
a) Shutting off the accumulator before maintenance
b) Discharging trapped fluid
c) Frequent replacement of nitrogen gas
d) Both (a) and (b)

15 | Review Questions

1. Define a hydraulic accumulator.
2. Explain the essential operation of a hydraulic accumulator using a simple circuit.
3. List four essential functions of hydraulic accumulators.
4. Explain how the pressure pulsations in a hydraulic rock drilling operation can be suppressed.
5. Describe the function of a hydraulic accumulator used as a shock absorber.
6. What adverse effect can happen in a primary hydraulic loader system with a 2-ton front-loading bucket if it is suddenly stopped? How can this effect be overcome?
7. Explain the function of a hydraulic accumulator used as a pump supplement.
8. Explain how the capacity requirements of the pump, utilized in a hydraulic system delivering only a short period of high-volume flow during a given cycle, can be reduced.
9. Develop a hydraulic circuit for carrying out a work process where significant fluid flows for a short period in each cycle.
10. Write briefly about the three most common applications of hydraulic accumulators.
11. Classify hydraulic accumulators.
12. Briefly explain the three basic types of accumulators used in hydraulic systems.
13. Draw the symbols for the weight-loaded, spring-loaded, and gas-charged hydraulic accumulators.

14. Explain the principle of operation of hydraulic accumulators.

15. Briefly explain the operation of a weight-loaded hydraulic accumulator, with a simple sketch.

16. Give some advantages and disadvantages of the weight-loaded hydraulic accumulators.

17. Briefly explain the operation of a spring-loaded hydraulic accumulator, with a simple sketch.

18. Give some benefits and disadvantages of the spring-loaded hydraulic accumulators.

19. Name three major classifications of the gas-charged hydraulic accumulators. Give one advantage of each accumulator.

20. Explain the fundamental operating principle of hydro-pneumatic accumulators.

21. What are the advantages of the gas-charged accumulators as compared to other types of hydraulic accumulators?

22. Why is the non-separator-type gas-charged accumulator not used in modern hydraulic systems?

23. Briefly explain the operation of a piston-type hydraulic accumulator, with a simple sketch.

24. Briefly explain the constructional features of piston-type hydraulic accumulators

25. Give one advantage and one disadvantage of the piston-type hydraulic accumulators.

26. Briefly explain the principle of operation of a hydraulic diaphragm accumulator, with a simple sketch.

27. Briefly describe the constructional features of the hydraulic diaphragm type accumulators

28. Give one advantage and one disadvantage of the hydraulic diaphragm accumulators.

29. Briefly explain the operation of a hydraulic bladder accumulator, with a simple sketch.

30. Explain the function and purpose of a gas-filled accumulator.

31. Briefly explain the constructional features of the hydraulic bladder-type accumulators

32. Give one advantage and one disadvantage of the hydraulic bladder accumulator.

33. Compare the diaphragm, bladder, and piston accumulators for typical sizes, pressures, flow rates, and suitability for shock suppression.

34. Give Short notes on (a) pre-charging the gas-loaded hydraulic accumulators and (b) the use of dump valves in hydraulic accumulator circuits.

35. What is the reason for connecting a throttle check valve at the inlet/outlet port of some accumulators?

36. Explain how a load-induced shock can be reduced in a hydraulic system, with a circuit diagram.

37. Explain the safety requirements of accumulators in hydraulic circuits.

38. Describe the functions of the safety-and-shut-off block associated with hydraulic accumulators.

39. Explain why it is required to unload an accumulator automatically when the machine is shut down.

40. Draw a hydraulic circuit for extending and retracting a double-acting hydraulic cylinder with provision for holding pressure, leakage compensation, and power savings. The circuit may include the following essential components: a pump, a check valve, an accumulator, a 4/3-DC valve, and a double-acting cylinder.

41. A large hydraulic cylinder is to clamp a workpiece. The cylinder requires a high volume of fluid as it extends rapidly. However, during the clamping period, the circuit requires no additional fluid. Assume that the circuit has an extended dwell time. Develop a circuit that is capable of storing energy at times of low demand and uses the stored

energy when a large volume of the fluid is required for a short period.

42. A diesel engine is started using a hydraulic motor. The maximum engine power requirement is only for a short time when the start signal is given with a 2/2-DC valve and the interval between two starting operations is long. An accumulator is used to supplement the pump's power during start-up. The pump should unload during idle time. Develop a hydraulic circuit to implement the scheme.

16 | Numerical Problems

1. Find the size of a gas-charged accumulator used to supply 1.95 gallons of fluid to a hydraulic cylinder with a maximum pressure of 2700 psi and a minimum pressure of 1827 psi. Assume that the accumulator is used for slow charging and slow discharging. The pre-charge pressure is 1740 psi. Assume negligible temperature variations.
[Ans: 6.36 gallons]

2. A gas-charged accumulator charges and discharges quickly at the rate of 200 in^3 in 3 seconds. It has a minimum to a maximum working pressure range of 1305 psi to 2611 psi and an operating temperature range of 68°F - 176°F. Assume the pre-charge pressure as 90% of the minimum pressure. Calculate the accumulator capacity.
[Ans: 2.9 gallons]

3. A gas-charged accumulator must discharge 281 in^3 of fluid in 3 seconds with a change of pressure from 4061 psi to 3017 psi. The loading time is 5 minutes. Define the capacity, keeping in mind that the ambient temperature changes from 680°F to 1760°F. Assume the adiabatic and isothermal correction factors for higher pressures are C_a = 0.75 and C_i = 0.75, respectively.

Objective-type questions: Answer key1-b, 2-b, 3-c, 4-b, 5-d

Appendix 1

Specifications of Piston Accumulators

Table A1.1 | Specifications of piston accumulators (1/3)

Gas capacity		Fluid capacity		Maximum working pressure	
cc	in^3	liter	Gallon	bar	psi
226	14	0.22	0.0625	345	5000
246	15	0.24	0.0625	207	3000
251	15	0.25	0.0625	690	10000
257	16	0.24	0.0625	172	2500
493	30	0.49	0.125	690	10000
498	30	0.49	0.125	345	5000
501	31	0.5	0.125	172	2500
524	32	0.5	0.125	207	3000
675	60	1	0.25	690	10000
991	61	1	0.25	172	2500
1000	61	1	0.25	345	5000
1016	62	1	0.25	207	3000
1900	116	2	0.5	345	5000
1901	116	2	0.5	207	3000
1910	117	2	0.5	690	10000
2163	132	2	0.5	172	2500
3048	186	3	0.75	345	5000
3868	236	4	1	345	5000
3966	242	4	1	207	3000
3992	244	4	1	690	10000
4080	249	4	1	172	2500
5779	353	6	1.5	345	5000
5867	358	6	1.5	207	3000
5914	361	6	1.5	690	10000

Table A1.1 | Specifications of piston accumulators (2/3)

Gas capacity		Fluid capacity		Maximum working pressure	
cc	in^3	liter	Gallon	bar	psi
5965	364	6	1.5	172	2500
7663	468	7	2	345	5000
7751	473	8	2	207	3000
7825	478	8	2	690	10000
7866	480	8	2	172	2500
9548	583	9	2.5	345	5000
9652	589	10	2.5	207	3000
9678	591	9	2.5	690	10000
9750	595	10	2.5	172	2500
11432	698	11	3	345	5000
11537	704	12	3	207	3000
11647	711	11	3	690	10000
15412	941	15	4	690	10000
19271	1176	19	5	207	3000
19292	1177	19	5	690	10000
19385	1183	19	5	345	5000
28743	1754	28	7.5	207	3000
28819	1759	28	7.5	345	5000
28849	1760	29	7.5	690	10000
38115	2326	38	10	690	10000
38198	2331	38	10	207	3000
38253	2334	37	10	345	5000
47687	2910	47	12.5	345	5000
47819	2918	47	12.5	207	3000
57054	3482	57	15	690	10000
57121	3486	56	15	345	5000
57266	3495	57	15	207	3000
66555	4061	66	17.5	345	5000
66713	4071	66	17.5	207	3000

Table A1.1 | Specifications of piston accumulators (3/3)

Gas capacity		Fluid capacity		Maximum working pressure	
cc	in^3	liter	Gallon	bar	psi
75989	4637	75	20	345	5000
75993	4637	76	20	690	10000
76160	4648	76	20	207	3000
90581	5528	86	23	207	3000
94938	5793	94	25	345	5000
95148	5806	95	25	207	3000
113806	6945	113	30	345	5000
114042	6959	114	30	207	3000
151379	9238	151	40	345	5000
151831	9265	151	40	207	3000
189619	11571	189	50	207	3000
191789	11704	189	50	345	5000
230246	14051	228	60	345	5000
231434	14123	227	60	207	3000
267312	16312	265	70	345	5000
269195	16427	265	70	207	3000
305537	18645	303	80	345	5000
306957	18732	303	80	207	3000
343298	20949	341	90	345	5000
344950	21050	341	90	207	3000
381292	23268	379	100	345	5000
382712	23355	379	100	207	3000
584045	35640	568	150	207	3000
773316	47190	757	200	207	3000

Appendix 2

Specifications of Bladder Accumulators

Table A2.1 | Specifications of bladder accumulators

Gas capacity		Fluid capacity		Maximum working pressure	
cc	in^3	liter	Gallon	bar	psi
1196	73	1	0.25	207	3000
3851	235	4	1	207	3000
9454	577	10	2.5	345	5000
9832	600	10	2.5	207	3000
18858	1151	19	5	345	5000
19714	1203	19	5	207	3000
35095	2142	38	10	345	5000
37018	2259	38	10	207	3000
41541	2535	42	11	207	3000
53413	3260	57	15	345	5000
56372	3440	57	15	207	3000

Appendix 3

Specifications of Diaphragm Accumulators

Table A3.1 | Specifications of diaphragm accumulators

Nominal volume		Maximum working pressure	
liter	Gallon	bar	psi
0.075	0.02	250	3625
0.16	0.04	210	3000
0.16	0.04	300	4350
0.32	0.08	100	1450
0.32	0.08	210	3000
0.32	0.08	300	4350
0.5	0.13	160	2320
0.5	0.13	210	3000
0.6	0.16	330	4780
0.6	0.16	350	5000
1	0.26	200	2900
1	0.26	330	4780
2	0.53	210	3000
2	0.53	250	3625
2	0.53	330	4780
2.8	0.74	210	3000
2.8	0.74	330	4780
3.5	0.92	210	3000
3.5	0.92	330	4780
4	1	50	725
4	1	180	2600

Appendix 4

A | Specification Parameters of Accumulators

Table 4.1 | Specifications parameters of accumulators (1/3)

Sl No	Parameter	Types/ selection / Examples (Typical)
1	Accumulator type	Piston Bladder Diaphragm (Weld type / Screw type) Bellows
2	Nominal volume	Piston: ½, 1 pint, 1, 2 quart, 1, 2.5, 4, 5, 7.5, 10, 12.5, 15, 20, 25, 30, 40, 50 gallons Bladder: ¼, 1, 2½, 5, 10, 14 gallons Diaphragm: 4.5, 10, 21, 30, 43, 85, 122, 170 in^3
3	Flow rate	Up to 3450 gpm
4	Effective gas volume	
5	Gas compression ratio	4:1, 6:1, 8:1, 10:1, unlimited
6	Maximum working pressure	1500, 2000, 3000, 3600, 4700, 6000, 7500, 10000, 15000, 20000 psi
7	Shell material	Carbon steel, stainless steel, carbon steel with a protective coating, low-temperature steel

Table 4.1 | Specifications parameters of accumulators (2/3)

Sl No	Parameter	Types/ Selection / Examples (Typical)
8	Diaphragm / Bladder / Seal material	Buna N (Acrylonitrile butadiene rubber) +5°F to +180°F Low temp Buna N (ECO) (Ethylene oxide epichlorohydrin rubber) -40°F to +180°F Viton (FKM) (Fluorine rubber) +5°F to +212°F Butyl rubber (IIR) -15°F to +80°F + EPDM, Polypropylene, Teflon, and Kel-F
9	Corrosion protection	Uncoated Coated Epoxy coated, Nickel/Chrome Plated (Inside surface)
10	Accumulator model (For bladder type)	Top repairable Bottom repairable
11	Fluid port size	As required
12	Fluid port material	Carbon steel, stainless steel, carbon steel with a protective coating, low-temperature steel

Table 4.1 | Specifications parameters of accumulators (3/3)

Sl No	Parameter	Types/ selection / Examples (Typical)
13	Fluid port connection	SAE O-ring Thread to ISO 228 (BSP) (G¾ to G2½) Thread to ISO 965/1 (Metric) Thread to DIN 13 Thread to ANSI B1.1 (UN.-2B seal SAE J 514) Thread to ANSI B1.20.1 (NPT) EN 1092-1 welding neck flange Flange ASME B16.5 SAE flange 3000 psi / 6000 psi
14	Gas port type	Sealed type / rechargeable type
15	Gas port size	As required
16	Gas port connection	M28 x1.5, M16x1.5, M14x1.5 M50x1.5 5/8- 18 UNF Not refillable (screw) Not refillable (welded)
17	Gas charge kit	As required
18	Operating temp	-50 to 350°F
19	Country of destination	As specified
20	Shell certification	ASME, TUV, PED (CE), CRN, AS1210
21	Mounting support	(e.g., Clamps and bracket)

B | Essential Specification Parameters, Safety Blocks

- Size (e.g., DN08, DN10, DN20, DN30)

- Maximum operating pressure (e.g., 5000 psi)

- Shut off valve, poppet type

- Bleed valve, poppet type (Manual, Manual + Solenoid)

- Pressure relief valve Rating (e.g., 5000 psi)

- Gauge

- Seal (e.g., Nitrile)

- Type of connection (e.g., Piping connection, sub-plate connection)

- Port sizes, G½ to G 1½

- Accumulator-side connection type (e.g., Threaded)

- Pressure port and gauge port connection types

- Tank port connection type

- Fluid compatibility (e.g., Mineral-based)

- Temperature rating

Appendix 5

Safety Standards for High-pressure Vessels

The design, manufacture, and use of the accumulator should conform to various national and international standards. Some of the standards of pressure vessels (accumulators) are given below.

Table 5.1 | Safety standards for high-pressure vessels

Standard	Applicable region or country
ASME BOILER and Pressure Vessel Code (American Society of Mechanical Engineers)	U S A
Pressure Equipment Directive - PED	Europe, including the UK
Reg. 365 B: Steam receivers, Separators, catch waters, Accumulators, and similar vessels under Indian Boiler Regulations	India
CSA B51, Boiler, pressure vessel, and pressure piping code, Canadian Registration Numbers (CRN)	Canada
Regulatory Rule NR-13, Brazilian Registered Engineers (BRE)	Brazil
AS1210, Pressure vessels, Standards Australia	Australia
High-Pressure Gas Safety Law (high-pressure gas production equipment)	Japan
Industrial Safety and Health Law (class-2 pressure vessel)	Japan
Regulation for Boiler and Pressure Vessel Manufacture Licensing	China
GOST standards	Russia
American Bureau of Shipping (ABS)	Shipping vessels and oil rigs
Off-shore Standard DNV-OS-E101, Det Norske Veritas (DNV)	Oil and gas applications

17 | References

1. Article on 'Advice For Maintaining Hydraulic Accumulators' by Brendan Casey, Machinery Lubrication (7/2009), by Noria

2. Article on: 'Hydraulic accumulators, Book 2, Chapter 1, Part 1, Part 2, Part 3', Hydraulics & Pneumatic Magazine.

3. Article on: 'What is an accumulator?' & TOBUL_Int_ Catalog_101912v2, Tobul Accumulator, Inc., USA.

4. Catalogue on: 'Accumulators V-FIFI-MC003-E' July 2005, EATON, Eden Prairie, MN, USA, www.hydraulics.eaton.com

5. Catalogue on: 'Sizing and Selection – Piston Accumulators, Bladder Accumulators, Kleen Vent', Catalog No. HY10-1630/US, Parker Hannifin Corporation, Hydraulic Accumulator Division, Illinois, USA.

6. Catalogues on 'Hydro-pneumatic accumulators', 'Precautions for use and maintenance recommendations', 'energy, silence, comfort, service life...', 'What are accumulators used for?', HYDRO LEDUC, FRANCE, www.hydroleduc.com

7. Document on 'Accumulators and Reservoirs', Senior Aerospace Metal Bellows, USA

8. Document on 'Accumulators', Bosch Rexroth Corporation, Industrial Hydraulics Division, USA

9. Document on 'Accumulators', Catalog V-FIFI-MC003-E July 2005, EATON Hydraulics, USA

10. Document on 'Bladder-type accumulator' RE 50170/02.12, Bosch Rexroth AG, Hydraulics, Germany

11. Document on 'Heavy Diesel Engines Metal Bellows Accumulators' HYDAC International, Germany

12. Document on: 'Accumulator sizing' EPE ITALIANA Srl, COLOGNO MONZESE (MI).

13. Document on: 'Accumulator station, RE 50135/07.11', Bosch Rexroth AG, Hydraulics, Germany

14. Publications Department of Womac Machine Supply Company, Fluid Power In Plant and Field, Second Edition
15. Vocational Training Course, HYDRAULICS – 21 Exercises with Instructions, published by Bundesinstitut fur Berufsbildungsforschung, Berlin, 1973

Fluid Power Educational Series Books

1. Pneumatic Systems and Circuits -Basic Level (In the SI Units)
2. Industrial Pneumatics -Basic Level (In the English Units)
3. Pneumatic Systems and Circuits -Advanced Level
4. Electro-Pneumatics and Automation
5. Design of Pneumatic Systems (In the SI Units)
6. Design Concepts in Pneumatic Systems (In the English Units)
7. Maintenance, Troubleshooting, and Safety in Pneumatic Systems
8. Industrial Hydraulic Systems and Circuits -Basic Level (In the SI Units)
9. Industrial Hydraulics -Basic Level (In the English Units)
10. Hydraulic Fluids
11. Hydraulic Filters: Construction, Installation Locations, and Specifications
12. Hydraulic Power Packs (In the SI Units)
13. Power Packs in Hydraulic Systems (In the English Units)
14. Hydraulic Cylinders (In the SI Units)
15. Hydraulic Linear Actuators (In the English Units)
16. Hydraulic Motors (In the SI Units)
17. Hydraulic Rotary Actuators (In the English Units)
18. Hydraulic Accumulators and Circuits (In the SI Units)
19. Accumulators in Hydraulic Systems (In the English Units)

For more details, please visit: https://jojibooks.com.

About the Author

Joji Parambath has been a trainer in Pneumatics, Hydraulics, and PLCs for over 25 years. During his career, he trained numerous professionals from the industries, as well as faculty members and students of engineering institutions.

At present, he is the key trainer at Fluidsys Training Centre, Bangalore, India (https://fluidsys.org), which provides training in Pneumatics and Hydraulics. He has already written two books on Pneumatics and Hydraulics. The publication of the present series of 36 books is intended to restructure and update the existing books.

The author wishes to thank all trainees for their lively interaction and many useful suggestions during the training programs that prompted the author to write the present series of books. You may send your feedback to info@fluidsys.org.

10th June 2020